MW00843585

Dear parents,

As a mom and as an educator, I am very
Workbook series with all of you. I developed this series for my two kids in elementary school, utilizing all of my knowledge and experience that I have gained while studying and working in the fields of Elementary Education and Gifted Education in South Korea as well as in the United States.

While raising my kids in the U.S., I had great disappointment and dissatisfaction about the math curriculum in the public schools. Based on my analysis, students cannot succeed in math with the current school curriculum because there is no sequential building up of fundamental skills. This is akin to building a castle on sand. So instead, I wanted to find a good workbook, but couldn't. And I also tried to find a tutor, but the price was too expensive for me. These are the reasons why I decided to make the Tiger Math series on my own.

The Tiger Math series was designed based on my three beliefs toward elementary math education.

1. It is extremely important to build foundation of math by acquiring a sense of numbers and mastering the four operation skills in terms of addition, subtraction, multiplication, and division.
2. In math, one should go through all steps in order, step by step, and cannot jump from level 1 to 3.
3. Practice math every day, even if only for 10 minutes.

If you feel that you don't know where your child should start, just choose a book in the Tiger Math series where your child thinks he/she can complete most of the material. And encourage your child to do only 2 sheets every day. When your child finishes the 2 sheets, review them together and encourage your child about his/her daily accomplishment.

I hope that the Tiger Math series can become a stepping stone for your child in gaining confidence and for making them interested in math as it has for my kids. Good luck!

Michelle Y. You, Ph.D.
Founder and CEO of Tiger Math

ACT scores show that only one out of four high school graduates are prepared to learn in college. This preparation needs to start early. In terms of basic math skills, being proficient in basic calculation means a lot. Help your child succeed by imparting basic math skills through hard work.

Sungwon S. Kim, Ph.D.
Engineering professor

Level D – 2: Plan of Study

Goal A	Practice subtracting 2 digit numbers from 2 digit numbers without borrowing. (Week 1)
Goal B	Practice subtracting 2 digit numbers from 2 digit numbers with borrowing. (Week 2 ~ 4)

Week 1

Day	Tiger Session		Topic	Goal
Mon	41	42	Subtraction: 2 digit – 2 digit	(10 ~ 99) - (10 ~ 99) without borrowing
Tue	43	44		
Wed	45	46		
Thu	47	48		
Fri	49	50		

Week 2

Day	Tiger Session		Topic	Goal
Mon	51	52	Subtraction: 2 digit – 2 digit	(10 ~ 99) - (10 ~ 99) with borrowing
Tue	53	54		
Wed	55	56		
Thu	57	58		
Fri	59	60		

Week 3

Day	Tiger Session		Topic	Goal
Mon	61	62	Subtraction: 2 digit – 2 digit	(10 ~ 99) - (10 ~ 99) with borrowing
Tue	63	64		
Wed	65	66		
Thu	67	68		
Fri	69	70		

Week 4

Day	Tiger Session		Topic	Goal
Mon	71	72	Subtraction: 2 digit – 2 digit	(10 ~ 99) - (10 ~ 99) with and without borrowing
Tue	73	74		
Wed	75	76		
Thu	77	78		
Fri	79	80		

Week 1

This week's goal is to practice subtracting 2 digit numbers from 2 digit numbers with without borrowing.

Tiger Session

Monday	41	42
Tuesday	43	44
Wednesday	45	46
Thursday	47	48
Friday	49	50

Date

Time spent | Score

min

41 2 digits − 2 digits ①

♠ **Subtract.**

Example

$$\begin{array}{r} 47 \\ -\ 13 \\ \hline \end{array} \quad \begin{array}{r} 47 \\ -\ 13 \\ \hline 4 \end{array} \quad \Rightarrow \quad \begin{array}{r} 47 \\ -\ 13 \\ \hline 34 \end{array}$$

(1) Tens Ones

$$\begin{array}{r} 38 \\ -\ 14 \\ \hline \end{array}$$

② ← ①

(2) Tens Ones

$$\begin{array}{r} 29 \\ -\ 12 \\ \hline \end{array}$$

② ← ①

(3)

$$\begin{array}{r} 31 \\ -\ 20 \\ \hline \end{array}$$

(4)

$$\begin{array}{r} 36 \\ -\ 13 \\ \hline \end{array}$$

(5)

$$\begin{array}{r} 35 \\ -\ 21 \\ \hline \end{array}$$

(6)

$$\begin{array}{r} 24 \\ -\ 12 \\ \hline \end{array}$$

(7) $\begin{array}{r} 22 \\ -\ 11 \\ \hline \end{array}$

(8) $\begin{array}{r} 27 \\ -\ 11 \\ \hline \end{array}$

(9) $\begin{array}{r} 39 \\ -\ 17 \\ \hline \end{array}$

(10) $\begin{array}{r} 33 \\ -\ 13 \\ \hline \end{array}$

(11) $\begin{array}{r} 38 \\ -\ 22 \\ \hline \end{array}$

(12) $\begin{array}{r} 25 \\ -\ 10 \\ \hline \end{array}$

(13) $\begin{array}{r} 37 \\ -\ 25 \\ \hline \end{array}$

(14) $\begin{array}{r} 39 \\ -\ 22 \\ \hline \end{array}$

(15) $\begin{array}{r} 26 \\ -\ 13 \\ \hline \end{array}$

(16) $\begin{array}{r} 29 \\ -\ 15 \\ \hline \end{array}$

42 2 digits − 2 digits ②

Date

Time spent: min

Score: /

♠ **Subtract.**

(1)

	Tens	Ones
	3	7
−	1	2
	②	← ①

(2)

	Tens	Ones
	3	6
−	2	2
	②	← ①

(3)
$$\begin{array}{r} 33 \\ -\ 12 \\ \hline \end{array}$$

(4)
$$\begin{array}{r} 38 \\ -\ 18 \\ \hline \end{array}$$

(5)
$$\begin{array}{r} 28 \\ -\ 12 \\ \hline \end{array}$$

(6)
$$\begin{array}{r} 36 \\ -\ 13 \\ \hline \end{array}$$

(7)
$$\begin{array}{r} 39 \\ -\ 26 \\ \hline \end{array}$$

(8)
$$\begin{array}{r} 22 \\ -\ 10 \\ \hline \end{array}$$

9) 23 students are in my class. My teacher decided to give a notebook for all students. But she mistakenly ordered only 12 notebooks. How many more notebooks does she need to order?

Equation: ______________________________

Answer: __________________

10) Mom bought 36 apples and 13 oranges. How many more apples did she buy than?

Equation: ______________________________

Answer: __________________

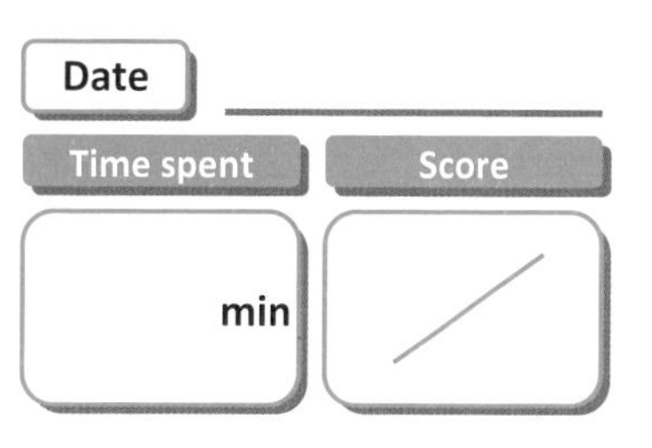

43 2 digits − 2 digits ③

♠ **Subtract.**

(1)

	Tens	Ones
	4	5
−	1	3
	② ←	①

(2)

	Tens	Ones
	3	3
−	1	2
	② ←	①

(3) $\begin{array}{r} 24 \\ -\ 21 \\ \hline \end{array}$

(4) $\begin{array}{r} 58 \\ -\ 14 \\ \hline \end{array}$

(5) $\begin{array}{r} 28 \\ -\ 14 \\ \hline \end{array}$

(6) $\begin{array}{r} 36 \\ -\ 15 \\ \hline \end{array}$

(7) $\begin{array}{r} 51 \\ -\ 21 \\ \hline \end{array}$

(8) $\begin{array}{r} 52 \\ -\ 11 \\ \hline \end{array}$

(9) $\begin{array}{r} 55 \\ -\ 15 \\ \hline \end{array}$

(10) $\begin{array}{r} 36 \\ -\ 12 \\ \hline \end{array}$

(11) $\begin{array}{r} 38 \\ -\ 28 \\ \hline \end{array}$

(12) $\begin{array}{r} 60 \\ -\ 10 \\ \hline \end{array}$

(13) $\begin{array}{r} 51 \\ -\ 11 \\ \hline \end{array}$

(14) $\begin{array}{r} 38 \\ -\ 25 \\ \hline \end{array}$

(15) $\begin{array}{r} 37 \\ -\ 23 \\ \hline \end{array}$

(16) $\begin{array}{r} 45 \\ -\ 14 \\ \hline \end{array}$

(17) $\begin{array}{r} 44 \\ -\ 12 \\ \hline \end{array}$

(18) $\begin{array}{r} 28 \\ -\ 16 \\ \hline \end{array}$

2 digits − 2 digits ④

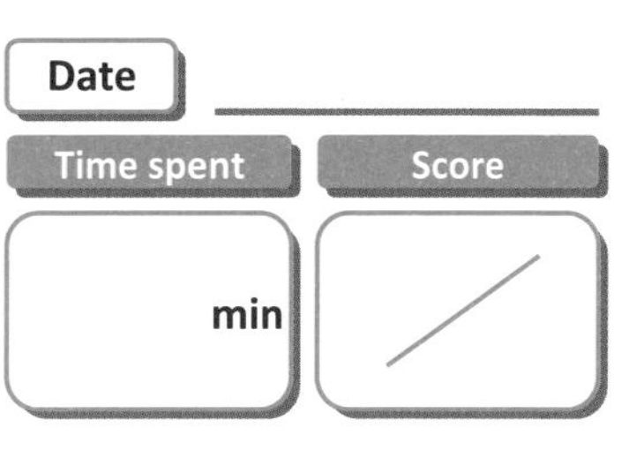

♠ **Subtract.**

(1)

	Tens	Ones
	5	6
−	1	5
	② ←	①

(2)

	Tens	Ones
	4	3
−	2	2
	② ←	①

(3)

$$\begin{array}{r} 33 \\ -\ 13 \\ \hline \end{array}$$

(4)

$$\begin{array}{r} 50 \\ -\ 20 \\ \hline \end{array}$$

(5)

$$\begin{array}{r} 59 \\ -\ 25 \\ \hline \end{array}$$

(6)

$$\begin{array}{r} 37 \\ -\ 22 \\ \hline \end{array}$$

(7)

$$\begin{array}{r} 27 \\ -\ 13 \\ \hline \end{array}$$

(8)

$$\begin{array}{r} 48 \\ -\ 12 \\ \hline \end{array}$$

9) In a toy shop, there are 32 robots and 56 dolls. How many more dolls are there in the toy shop?

Equation: ______________________________

Answer: ___________________

10) In a toy shop, there were 48 robots. If 12 robots are sold, then how many robots are left?

Equation: ______________________________

Answer: ___________________

Date

Time spent | Score

min

45 2 digits − 2 digits ⑤

♠ **Subtract.**

(1)

	Tens	Ones
	6	4
−	1	1

② ← ①

(2)

	Tens	Ones
	5	9
−	1	2

② ← ①

(3) $47 - 22$

(4) $67 - 23$

(5) $55 - 41$

(6) $73 - 41$

(7) $46 - 24$

(8) $78 - 35$

(9) $\begin{array}{r} 45 \\ -\ 25 \\ \hline \end{array}$

(10) $\begin{array}{r} 58 \\ -\ 36 \\ \hline \end{array}$

(11) $\begin{array}{r} 54 \\ -\ 31 \\ \hline \end{array}$

(12) $\begin{array}{r} 76 \\ -\ 15 \\ \hline \end{array}$

(13) $\begin{array}{r} 68 \\ -\ 20 \\ \hline \end{array}$

(14) $\begin{array}{r} 49 \\ -\ 15 \\ \hline \end{array}$

(15) $\begin{array}{r} 54 \\ -\ 41 \\ \hline \end{array}$

(16) $\begin{array}{r} 67 \\ -\ 17 \\ \hline \end{array}$

(17) $\begin{array}{r} 70 \\ -\ 30 \\ \hline \end{array}$

(18) $\begin{array}{r} 55 \\ -\ 41 \\ \hline \end{array}$

2 digits − 2 digits ⑥

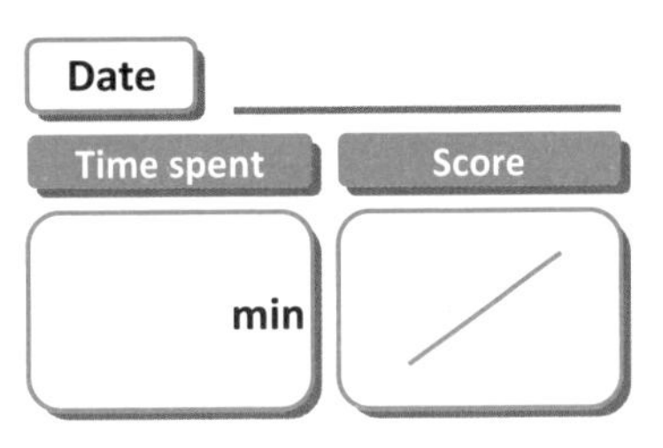

♠ **Subtract.**

(1)

	Tens	Ones
	5	5
−	4	4
	② ← ①	

(2)

	Tens	Ones
	6	7
−	2	7
	② ← ①	

(3) $\begin{array}{r} 73 \\ -\ 22 \\ \hline \end{array}$

(4) $\begin{array}{r} 49 \\ -\ 33 \\ \hline \end{array}$

(5) $\begin{array}{r} 57 \\ -\ 34 \\ \hline \end{array}$

(6) $\begin{array}{r} 70 \\ -\ 10 \\ \hline \end{array}$

(7) $\begin{array}{r} 62 \\ -\ 31 \\ \hline \end{array}$

(8) $\begin{array}{r} 58 \\ -\ 44 \\ \hline \end{array}$

9) John jump-roped 56 times yesterday and 77 times today. How many more times did he jump-rope today?

Equation: ______________________

Answer: ______________

10) There were 65 pencils in a box. If there are 32 pencils left in the box two month later, how many pencils were used?

Equation: ______________________

Answer: ______________

Date

Time spent | Score

min

47 2 digits − 2 digits ⑦

♠ **Subtract.**

(1)

	Tens	Ones
	4	6
−	3	3
	② ←	①

(2)

	Tens	Ones
	9	5
−	6	4
	② ←	①

(3) $\begin{array}{r} 77 \\ -\ 12 \\ \hline \end{array}$

(4) $\begin{array}{r} 52 \\ -\ 20 \\ \hline \end{array}$

(5) $\begin{array}{r} 83 \\ -\ 31 \\ \hline \end{array}$

(6) $\begin{array}{r} 57 \\ -\ 11 \\ \hline \end{array}$

(7) $\begin{array}{r} 59 \\ -\ 15 \\ \hline \end{array}$

(8) $\begin{array}{r} 68 \\ -\ 31 \\ \hline \end{array}$

(9) $\begin{array}{r} 43 \\ -\ 32 \\ \hline \end{array}$

(10) $\begin{array}{r} 59 \\ -\ 12 \\ \hline \end{array}$

(11) $\begin{array}{r} 64 \\ -\ 24 \\ \hline \end{array}$

(12) $\begin{array}{r} 76 \\ -\ 54 \\ \hline \end{array}$

(13) $\begin{array}{r} 45 \\ -\ 22 \\ \hline \end{array}$

(14) $\begin{array}{r} 98 \\ -\ 63 \\ \hline \end{array}$

(15) $\begin{array}{r} 89 \\ -\ 48 \\ \hline \end{array}$

(16) $\begin{array}{r} 70 \\ -\ 10 \\ \hline \end{array}$

(17) $\begin{array}{r} 53 \\ -\ 33 \\ \hline \end{array}$

(18) $\begin{array}{r} 57 \\ -\ 31 \\ \hline \end{array}$

48 2 digits − 2 digits ⑧

Date ____

Time spent: min | Score: /

♠ **Subtract.**

(1)
$$\begin{array}{r} 53 \\ -\ 20 \\ \hline \end{array}$$
Tens | Ones — ② ← ①

(2)
$$\begin{array}{r} 67 \\ -\ 42 \\ \hline \end{array}$$
Tens | Ones — ② ← ①

(3)
$$\begin{array}{r} 99 \\ -\ 71 \\ \hline \end{array}$$

(4)
$$\begin{array}{r} 80 \\ -\ 50 \\ \hline \end{array}$$

(5)
$$\begin{array}{r} 46 \\ -\ 15 \\ \hline \end{array}$$

(6)
$$\begin{array}{r} 86 \\ -\ 63 \\ \hline \end{array}$$

(7)
$$\begin{array}{r} 76 \\ -\ 52 \\ \hline \end{array}$$

(8)
$$\begin{array}{r} 54 \\ -\ 43 \\ \hline \end{array}$$

9) Katy runs a pizzeria. She sold 56 pizzas yesterday and 98 pizzas today. How many more pizzas did she sell today compared to yesterday?

Equation: ______________________________

Answer: __________________

10) Liam swam 45 laps, and Chris swam 57 laps today. How many more laps did Chris swim than Liam?

Equation: ______________________________

Answer: __________________

Date ____

Time spent: ____ min | Score: ____

49 2 digits − 2 digits ⑨

♠ **Subtract.**

(1)
	Tens	Ones
	4	3
−	2	2

② ← ①

(2)
	Tens	Ones
	7	5
−	3	1

② ← ①

(3) $\begin{array}{r} 99 \\ -\ 65 \\ \hline \end{array}$

(4) $\begin{array}{r} 35 \\ -\ 12 \\ \hline \end{array}$

(5) $\begin{array}{r} 44 \\ -\ 22 \\ \hline \end{array}$

(6) $\begin{array}{r} 67 \\ -\ 42 \\ \hline \end{array}$

(7) $\begin{array}{r} 86 \\ -\ 50 \\ \hline \end{array}$

(8) $\begin{array}{r} 58 \\ -\ 44 \\ \hline \end{array}$

(9) $\begin{array}{r} 88 \\ -\ 42 \\ \hline \end{array}$

(10) $\begin{array}{r} 36 \\ -\ 23 \\ \hline \end{array}$

(11) $\begin{array}{r} 45 \\ -\ 14 \\ \hline \end{array}$

(12) $\begin{array}{r} 97 \\ -\ 52 \\ \hline \end{array}$

(13) $\begin{array}{r} 84 \\ -\ 34 \\ \hline \end{array}$

(14) $\begin{array}{r} 67 \\ -\ 45 \\ \hline \end{array}$

(15) $\begin{array}{r} 56 \\ -\ 23 \\ \hline \end{array}$

(16) $\begin{array}{r} 46 \\ -\ 31 \\ \hline \end{array}$

(17) $\begin{array}{r} 51 \\ -\ 20 \\ \hline \end{array}$

(18) $\begin{array}{r} 94 \\ -\ 51 \\ \hline \end{array}$

Date

Time spent | Score

min

50 2 digits − 2 digits ⑩

♠ **Subtract.**

(1)
	Tens	Ones
	6	8
−	2	0
	② ← ①	

(2)
	Tens	Ones
	9	5
−	4	2
	② ← ①	

(3) $\begin{array}{r} 46 \\ -\ 33 \\ \hline \end{array}$

(4) $\begin{array}{r} 36 \\ -\ 25 \\ \hline \end{array}$

(5) $\begin{array}{r} 58 \\ -\ 43 \\ \hline \end{array}$

(6) $\begin{array}{r} 78 \\ -\ 56 \\ \hline \end{array}$

(7) $\begin{array}{r} 83 \\ -\ 51 \\ \hline \end{array}$

(8) $\begin{array}{r} 54 \\ -\ 21 \\ \hline \end{array}$

9) In a parking lot, there are 22 trucks and 65 vans. How many more vans are there in the parking lot?

Equation: ______________________________

Answer: __________________

10) My grandma is 77 years old, and my mom is 37 years old. How many years is my grandma older than my mom?

Equation: ______________________________

Answer: __________________

Week 2

This week's goal is to practice subtracting 2 digit numbers from 2 digit numbers with borrowing.

Tiger Session

Monday	51	52
Tuesday	53	54
Wednesday	55	56
Thursday	57	58
Friday	59	60

51 2 digits − 2 digits ⑪

Date ______

Time spent: min

Score:

♠ **Subtract.**

Example

$$\begin{array}{r} 53 \\ -\,16 \\ \hline \end{array}$$

Oops! You can't subtract 6 from 3! That's ok. You can **borrow 10** from the tens digit. Let's practice below.

$$\begin{array}{r} 53 \\ -\,16 \\ \hline \end{array} \rightarrow \begin{array}{r} \overset{4}{\cancel{5}}\,3 \\ -\,1\,6 \\ \hline \end{array} \rightarrow \begin{array}{r} \overset{4}{\cancel{5}}\,\overset{13}{\cancel{3}} \\ -\,1\,6 \\ \hline \end{array}$$

$$\rightarrow \begin{array}{r} \overset{4}{\cancel{5}}\,\overset{13}{\cancel{3}} \\ -\,1\,6 \\ \hline 7 \end{array} \rightarrow \begin{array}{r} \overset{4}{\cancel{5}}\,\overset{13}{\cancel{3}} \\ -\,1\,6 \\ \hline 3\,7 \end{array}$$

(1)
$$\begin{array}{r} \square\,\square \\ 3\,7 \\ -\,1\,8 \\ \hline \end{array}$$

(2)
$$\begin{array}{r} \square\,\square \\ 2\,2 \\ -\,1\,9 \\ \hline \end{array}$$

(3)
$$\begin{array}{r} \square\square \\ 31 \\ -\ 16 \\ \hline \end{array}$$

(4)
$$\begin{array}{r} \square\square \\ 32 \\ -\ 18 \\ \hline \end{array}$$

(5)
$$\begin{array}{r} \square\square \\ 35 \\ -\ 19 \\ \hline \end{array}$$

(6)
$$\begin{array}{r} \square\square \\ 24 \\ -\ 16 \\ \hline \end{array}$$

(7)
$$\begin{array}{r} \square\square \\ 23 \\ -\ 14 \\ \hline \end{array}$$

(8)
$$\begin{array}{r} \square\square \\ 27 \\ -\ \ 9 \\ \hline \end{array}$$

(9)
$$\begin{array}{r} \square\square \\ 30 \\ -\ 15 \\ \hline \end{array}$$

(10)
$$\begin{array}{r} \square\square \\ 25 \\ -\ 18 \\ \hline \end{array}$$

Date

Time spent | Score

min

52 2 digits − 2 digits ⑫

♠ **Subtract.**

(1) $\begin{array}{r} 22 \\ -\ 16 \\ \hline \end{array}$

(2) $\begin{array}{r} 27 \\ -\ 18 \\ \hline \end{array}$

(3) $\begin{array}{r} 34 \\ -\ 17 \\ \hline \end{array}$

(4) $\begin{array}{r} 33 \\ -\ 18 \\ \hline \end{array}$

(5) $\begin{array}{r} 31 \\ -\ 17 \\ \hline \end{array}$

(6) $\begin{array}{r} 21 \\ -\ 15 \\ \hline \end{array}$

(7) $\begin{array}{r} 37 \\ -\ 19 \\ \hline \end{array}$

(8) $\begin{array}{r} 32 \\ -\ 28 \\ \hline \end{array}$

9) My mom is 29 years old, and my dad is 35 years old. How many years is my dad older than my mom?

Equation: ______________________

Answer: ______________

10) Christine had 33 pebbles. One day she gave 18 out of the 33 to her friends as gifts. How many pebbles does she have left?

Equation: ______________________

Answer: ______________

53 2 digits − 2 digits ⑬

Date

Time spent: min | Score: /

♠ **Subtract.**

(1)
$$\begin{array}{r} 33 \\ -\ 17 \\ \hline \end{array}$$

(2)
$$\begin{array}{r} 34 \\ -\ 18 \\ \hline \end{array}$$

(3)
$$\begin{array}{r} 33 \\ -\ 15 \\ \hline \end{array}$$

(4)
$$\begin{array}{r} 35 \\ -\ 16 \\ \hline \end{array}$$

(5)
$$\begin{array}{r} 23 \\ -\ 14 \\ \hline \end{array}$$

(6)
$$\begin{array}{r} 22 \\ -\ 18 \\ \hline \end{array}$$

(7)
$$\begin{array}{r} 26 \\ -\ 19 \\ \hline \end{array}$$

(8)
$$\begin{array}{r} 31 \\ -\ 15 \\ \hline \end{array}$$

(9) $\begin{array}{r} \square\square \\ 21 \\ -\ 18 \\ \hline \end{array}$

(10) $\begin{array}{r} \square\square \\ 35 \\ -\ 27 \\ \hline \end{array}$

(11) $\begin{array}{r} \square\square \\ 24 \\ -\ 19 \\ \hline \end{array}$

(12) $\begin{array}{r} \square\square \\ 28 \\ -\ 19 \\ \hline \end{array}$

(13) $\begin{array}{r} \square\square \\ 23 \\ -\ 14 \\ \hline \end{array}$

(14) $\begin{array}{r} \square\square \\ 36 \\ -\ 19 \\ \hline \end{array}$

(15) $\begin{array}{r} \square\square \\ 31 \\ -\ 17 \\ \hline \end{array}$

(16) $\begin{array}{r} \square\square \\ 25 \\ -\ 18 \\ \hline \end{array}$

(17) $\begin{array}{r} \square\square \\ 32 \\ -\ 16 \\ \hline \end{array}$

(18) $\begin{array}{r} \square\square \\ 37 \\ -\ 19 \\ \hline \end{array}$

Date

Time spent | Score

min

54 2 digits − 2 digits ⑭

♠ **Subtract.**

(1)
$$\begin{array}{r} 25 \\ -\ 17 \\ \hline \end{array}$$

(2)
$$\begin{array}{r} 36 \\ -\ 29 \\ \hline \end{array}$$

(3)
$$\begin{array}{r} 38 \\ -\ 19 \\ \hline \end{array}$$

(4)
$$\begin{array}{r} 30 \\ -\ 15 \\ \hline \end{array}$$

(5)
$$\begin{array}{r} 21 \\ -\ 15 \\ \hline \end{array}$$

(6)
$$\begin{array}{r} 38 \\ -\ 19 \\ \hline \end{array}$$

(7)
$$\begin{array}{r} 37 \\ -\ 19 \\ \hline \end{array}$$

(8)
$$\begin{array}{r} 22 \\ -\ 18 \\ \hline \end{array}$$

9) Gavin ran 19 laps, and his mom ran 36 laps. How many more laps did his mom run?

Equation: ______________________

Answer: ______________

10) Dad bought 24 bottles of water at a store and brought them home. After a week, we found that 15 bottles are left. How many bottles of water did my family drink?

Equation: ______________________

Answer: ______________

55 2 digits − 2 digits ⑮

Date

Time spent | Score
min

♠ **Subtract.**

(1)
$$\begin{array}{r} 44 \\ -\ 15 \\ \hline \end{array}$$

(2)
$$\begin{array}{r} 50 \\ -\ 17 \\ \hline \end{array}$$

(3)
$$\begin{array}{r} 47 \\ -\ 28 \\ \hline \end{array}$$

(4)
$$\begin{array}{r} 57 \\ -\ 29 \\ \hline \end{array}$$

(5)
$$\begin{array}{r} 51 \\ -\ 33 \\ \hline \end{array}$$

(6)
$$\begin{array}{r} 40 \\ -\ 24 \\ \hline \end{array}$$

(7)
$$\begin{array}{r} 46 \\ -\ 18 \\ \hline \end{array}$$

(8)
$$\begin{array}{r} 48 \\ -\ 29 \\ \hline \end{array}$$

(9)
$$\begin{array}{r} \square\square \\ 40 \\ -\ 25 \\ \hline \end{array}$$

(10)
$$\begin{array}{r} \square\square \\ 58 \\ -\ 29 \\ \hline \end{array}$$

(11)
$$\begin{array}{r} \square\square \\ 54 \\ -\ 18 \\ \hline \end{array}$$

(12)
$$\begin{array}{r} \square\square \\ 50 \\ -\ 17 \\ \hline \end{array}$$

(13)
$$\begin{array}{r} \square\square \\ 51 \\ -\ 27 \\ \hline \end{array}$$

(14)
$$\begin{array}{r} \square\square \\ 41 \\ -\ 26 \\ \hline \end{array}$$

(15)
$$\begin{array}{r} \square\square \\ 54 \\ -\ 18 \\ \hline \end{array}$$

(16)
$$\begin{array}{r} \square\square \\ 57 \\ -\ 29 \\ \hline \end{array}$$

(17)
$$\begin{array}{r} \square\square \\ 40 \\ -\ 17 \\ \hline \end{array}$$

(18)
$$\begin{array}{r} \square\square \\ 55 \\ -\ 26 \\ \hline \end{array}$$

56 2 digits − 2 digits ⑯

Date ______

Time spent ____ min | Score ____

♠ **Subtract.**

(1)
$$\begin{array}{r} 55 \\ -\ 37 \\ \hline \end{array}$$

(2)
$$\begin{array}{r} 41 \\ -\ 16 \\ \hline \end{array}$$

(3)
$$\begin{array}{r} 41 \\ -\ 12 \\ \hline \end{array}$$

(4)
$$\begin{array}{r} 46 \\ -\ 29 \\ \hline \end{array}$$

(5)
$$\begin{array}{r} 47 \\ -\ 38 \\ \hline \end{array}$$

(6)
$$\begin{array}{r} 40 \\ -\ 19 \\ \hline \end{array}$$

(7)
$$\begin{array}{r} 42 \\ -\ 27 \\ \hline \end{array}$$

(8)
$$\begin{array}{r} 58 \\ -\ 19 \\ \hline \end{array}$$

9) There are 52 pants and 29 t-shirts on display in a store. How many more pants are on display?

Equation: ______________________

Answer: ______________

10) Gavin collected 45 shells on the beach. Coming back home, he gave 26 shells to his sister. How many shells does he have now?

Equation: ______________________

Answer: ______________

Date ______

Time spent | Score
min |

57 2 digits − 2 digits ⑰

♠ **Subtract.**

(1)
$$\begin{array}{r} 46 \\ -\ 18 \\ \hline \end{array}$$

(2)
$$\begin{array}{r} 40 \\ -\ 23 \\ \hline \end{array}$$

(3)
$$\begin{array}{r} 47 \\ -\ 29 \\ \hline \end{array}$$

(4)
$$\begin{array}{r} 41 \\ -\ 18 \\ \hline \end{array}$$

(5)
$$\begin{array}{r} 43 \\ -\ 17 \\ \hline \end{array}$$

(6)
$$\begin{array}{r} 47 \\ -\ 19 \\ \hline \end{array}$$

(7)
$$\begin{array}{r} 57 \\ -\ 18 \\ \hline \end{array}$$

(8)
$$\begin{array}{r} 44 \\ -\ 29 \\ \hline \end{array}$$

(9) $\begin{array}{r} \square\square \\ 42 \\ -\ 15 \\ \hline \end{array}$

(10) $\begin{array}{r} \square\square \\ 43 \\ -\ 16 \\ \hline \end{array}$

(11) $\begin{array}{r} \square\square \\ 41 \\ -\ 17 \\ \hline \end{array}$

(12) $\begin{array}{r} \square\square \\ 46 \\ -\ 28 \\ \hline \end{array}$

(13) $\begin{array}{r} \square\square \\ 43 \\ -\ 29 \\ \hline \end{array}$

(14) $\begin{array}{r} \square\square \\ 58 \\ -\ 39 \\ \hline \end{array}$

(15) $\begin{array}{r} \square\square \\ 45 \\ -\ 38 \\ \hline \end{array}$

(16) $\begin{array}{r} \square\square \\ 50 \\ -\ 31 \\ \hline \end{array}$

(17) $\begin{array}{r} \square\square \\ 53 \\ -\ 36 \\ \hline \end{array}$

(18) $\begin{array}{r} \square\square \\ 47 \\ -\ 29 \\ \hline \end{array}$

Date

Time spent | Score

min

58 2 digits − 2 digits ⑱

♠ **Subtract.**

(1) $\begin{array}{r} 53 \\ -\ 39 \\ \hline \end{array}$

(2) $\begin{array}{r} 47 \\ -\ 28 \\ \hline \end{array}$

(3) $\begin{array}{r} 55 \\ -\ 29 \\ \hline \end{array}$

(4) $\begin{array}{r} 50 \\ -\ 37 \\ \hline \end{array}$

(5) $\begin{array}{r} 46 \\ -\ 18 \\ \hline \end{array}$

(6) $\begin{array}{r} 54 \\ -\ 27 \\ \hline \end{array}$

(7) $\begin{array}{r} 45 \\ -\ 29 \\ \hline \end{array}$

(8) $\begin{array}{r} 44 \\ -\ 19 \\ \hline \end{array}$

9) 32 kittens and 26 puppies are playing in a pet shop. How many more kittens are there?

Equation: ______________________________

Answer: __________________

10) Yesterday, 34 kids visited the library, and today 52 kids visited. How many more kids visited the library today than yesterday?

Equation: ______________________________

Answer: __________________

59 2 digits − 2 digits ⑲

Date

Time spent min

Score

♠ **Subtract.**

(1) $\begin{array}{r} 25 \\ -\ 16 \\ \hline \end{array}$

(2) $\begin{array}{r} 21 \\ -\ 16 \\ \hline \end{array}$

(3) $\begin{array}{r} 47 \\ -\ 18 \\ \hline \end{array}$

(4) $\begin{array}{r} 25 \\ -\ 19 \\ \hline \end{array}$

(5) $\begin{array}{r} 44 \\ -\ 15 \\ \hline \end{array}$

(6) $\begin{array}{r} 33 \\ -\ 16 \\ \hline \end{array}$

(7) $\begin{array}{r} 40 \\ -\ 26 \\ \hline \end{array}$

(8) $\begin{array}{r} 24 \\ -\ 17 \\ \hline \end{array}$

(9)
$$\begin{array}{r} \square\square \\ 58 \\ -\ 19 \\ \hline \end{array}$$

(10)
$$\begin{array}{r} \square\square \\ 20 \\ -\ 12 \\ \hline \end{array}$$

(11)
$$\begin{array}{r} \square\square \\ 24 \\ -\ 15 \\ \hline \end{array}$$

(12)
$$\begin{array}{r} \square\square \\ 36 \\ -\ 19 \\ \hline \end{array}$$

(13)
$$\begin{array}{r} \square\square \\ 50 \\ -\ 11 \\ \hline \end{array}$$

(14)
$$\begin{array}{r} \square\square \\ 23 \\ -\ 17 \\ \hline \end{array}$$

(15)
$$\begin{array}{r} \square\square \\ 45 \\ -\ 19 \\ \hline \end{array}$$

(16)
$$\begin{array}{r} \square\square \\ 34 \\ -\ 15 \\ \hline \end{array}$$

(17)
$$\begin{array}{r} \square\square \\ 36 \\ -\ 19 \\ \hline \end{array}$$

(18)
$$\begin{array}{r} \square\square \\ 27 \\ -\ 18 \\ \hline \end{array}$$

2 digits − 2 digits ⑳

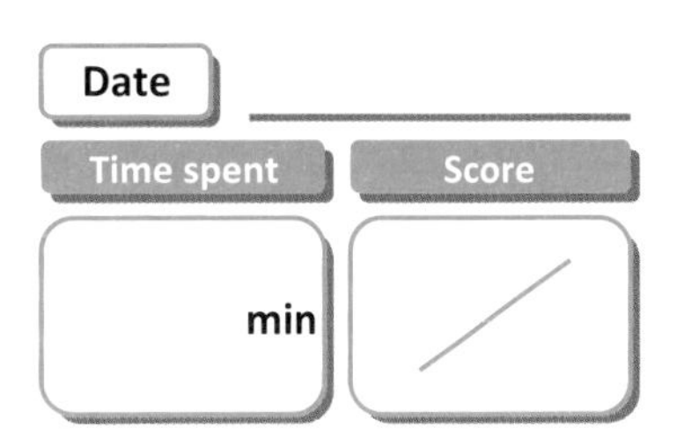

♠ **Subtract.**

(1)
$$\begin{array}{r} 31 \\ -\ 16 \\ \hline \end{array}$$

(2)
$$\begin{array}{r} 54 \\ -\ 27 \\ \hline \end{array}$$

(3)
$$\begin{array}{r} 35 \\ -\ 28 \\ \hline \end{array}$$

(4)
$$\begin{array}{r} 36 \\ -\ 17 \\ \hline \end{array}$$

(5)
$$\begin{array}{r} 52 \\ -\ 39 \\ \hline \end{array}$$

(6)
$$\begin{array}{r} 40 \\ -\ 28 \\ \hline \end{array}$$

(7)
$$\begin{array}{r} 36 \\ -\ 18 \\ \hline \end{array}$$

(8)
$$\begin{array}{r} 46 \\ -\ 19 \\ \hline \end{array}$$

9) Ryan brought 42 pieces of candy to the school. If he gave 35 pieces of candy to his classmates, how many pieces are left?

Equation: ______________________________

Answer: __________________

10) In my class, the teacher gives stickers to students who behave well that day. When a student collects 30 stickers, he or she becomes a superstar. So far, I have collected 15 stickers. How many more stickers do I have to collect to become a superstar?

Equation: ______________________________

Answer: __________________

Week 3

This week's goal is to practice subtracting 2 digit numbers from 2 digit numbers with borrowing.

Tiger Session

Day		
Monday	61	62
Tuesday	63	64
Wednesday	65	66
Thursday	67	68
Friday	69	70

61 2 digits − 2 digits ㉑

Date ____

Time spent ____ min

Score ____ /

♠ **Subtract.**

(1) $\begin{array}{r} \square\square \\ 64 \\ -\ 26 \\ \hline \end{array}$

(2) $\begin{array}{r} \square\square \\ 77 \\ -\ 28 \\ \hline \end{array}$

(3) $\begin{array}{r} 60 \\ -\ 34 \\ \hline \end{array}$

(4) $\begin{array}{r} 62 \\ -\ 46 \\ \hline \end{array}$

(5) $\begin{array}{r} 64 \\ -\ 47 \\ \hline \end{array}$

(6) $\begin{array}{r} 62 \\ -\ 38 \\ \hline \end{array}$

(7) $\begin{array}{r} 71 \\ -\ 26 \\ \hline \end{array}$

(8) $\begin{array}{r} 72 \\ -\ 19 \\ \hline \end{array}$

(9) $\begin{array}{r} 64 \\ -\ 28 \\ \hline \end{array}$

(10) $\begin{array}{r} 75 \\ -\ 49 \\ \hline \end{array}$

(11) $\begin{array}{r} 68 \\ -\ 26 \\ \hline \end{array}$

(12) $\begin{array}{r} 60 \\ -\ 22 \\ \hline \end{array}$

(13) $\begin{array}{r} 70 \\ -\ 55 \\ \hline \end{array}$

(14) $\begin{array}{r} 68 \\ -\ 39 \\ \hline \end{array}$

(15) $\begin{array}{r} 65 \\ -\ 19 \\ \hline \end{array}$

(16) $\begin{array}{r} 79 \\ -\ 24 \\ \hline \end{array}$

(17) $\begin{array}{r} 75 \\ -\ 18 \\ \hline \end{array}$

(18) $\begin{array}{r} 73 \\ -\ 25 \\ \hline \end{array}$

62 2 digits − 2 digits ㉒

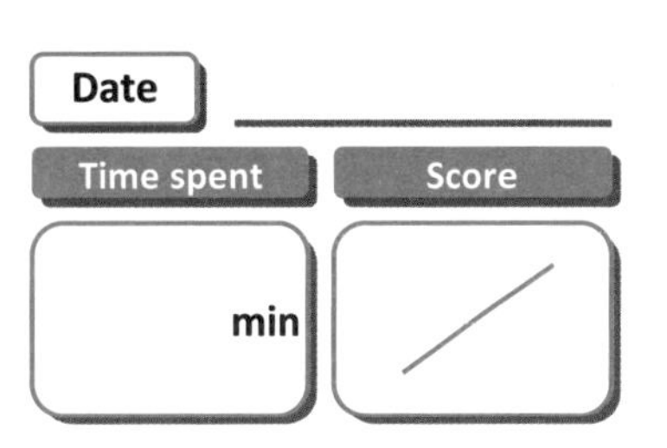

♠ **Subtract.**

(1)
$$\begin{array}{r} 65 \\ -\ 28 \\ \hline \end{array}$$

(2)
$$\begin{array}{r} 74 \\ -\ 17 \\ \hline \end{array}$$

(3)
$$\begin{array}{r} 63 \\ -\ 55 \\ \hline \end{array}$$

(4)
$$\begin{array}{r} 61 \\ -\ 43 \\ \hline \end{array}$$

(5)
$$\begin{array}{r} 64 \\ -\ 29 \\ \hline \end{array}$$

(6)
$$\begin{array}{r} 75 \\ -\ 49 \\ \hline \end{array}$$

(7)
$$\begin{array}{r} 68 \\ -\ 19 \\ \hline \end{array}$$

(8)
$$\begin{array}{r} 72 \\ -\ 27 \\ \hline \end{array}$$

9) In 3rd grade in my school, there are 68 girls and 73 boys. How many more boys are there than girls?

Equation: ______________________

Answer: ______________

10) The teacher asked students in the 3rd grade what their favorite sport was. 45 students answered basketball and 61 said football. How many more students answered that football was their favorite sport?

Equation: ______________________

Answer: ______________

Date

Time spent | Score

min

63 2 digits − 2 digits ㉓

♠ **Subtract.**

(1)
$$\begin{array}{r} 60 \\ -\ 35 \\ \hline \end{array}$$

(2)
$$\begin{array}{r} 61 \\ -\ 19 \\ \hline \end{array}$$

(3)
$$\begin{array}{r} 68 \\ -\ 29 \\ \hline \end{array}$$

(4)
$$\begin{array}{r} 73 \\ -\ 36 \\ \hline \end{array}$$

(5)
$$\begin{array}{r} 73 \\ -\ 44 \\ \hline \end{array}$$

(6)
$$\begin{array}{r} 61 \\ -\ 25 \\ \hline \end{array}$$

(7)
$$\begin{array}{r} 67 \\ -\ 48 \\ \hline \end{array}$$

(8)
$$\begin{array}{r} 60 \\ -\ 18 \\ \hline \end{array}$$

(9) $\begin{array}{r} 62 \\ -\ 18 \\ \hline \end{array}$

(10) $\begin{array}{r} 75 \\ -\ 38 \\ \hline \end{array}$

(11) $\begin{array}{r} 64 \\ -\ 28 \\ \hline \end{array}$

(12) $\begin{array}{r} 60 \\ -\ 25 \\ \hline \end{array}$

(13) $\begin{array}{r} 71 \\ -\ 49 \\ \hline \end{array}$

(14) $\begin{array}{r} 70 \\ -\ 59 \\ \hline \end{array}$

(15) $\begin{array}{r} 62 \\ -\ 39 \\ \hline \end{array}$

(16) $\begin{array}{r} 66 \\ -\ 27 \\ \hline \end{array}$

(17) $\begin{array}{r} 61 \\ -\ 36 \\ \hline \end{array}$

(18) $\begin{array}{r} 65 \\ -\ 47 \\ \hline \end{array}$

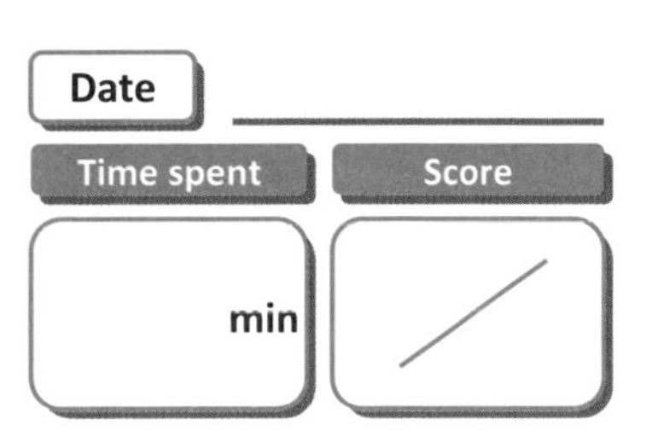

2 digits − 2 digits ㉔

♠ **Subtract.**

(1)
$$\begin{array}{r} \square\,\square \\ 76 \\ -\ 28 \\ \hline \end{array}$$

(2)
$$\begin{array}{r} \square\,\square \\ 60 \\ -\ 36 \\ \hline \end{array}$$

(3)
$$\begin{array}{r} 78 \\ -\ 39 \\ \hline \end{array}$$

(4)
$$\begin{array}{r} 64 \\ -\ 28 \\ \hline \end{array}$$

(5)
$$\begin{array}{r} 62 \\ -\ 29 \\ \hline \end{array}$$

(6)
$$\begin{array}{r} 72 \\ -\ 58 \\ \hline \end{array}$$

(7)
$$\begin{array}{r} 67 \\ -\ 49 \\ \hline \end{array}$$

(8)
$$\begin{array}{r} 75 \\ -\ 29 \\ \hline \end{array}$$

9) In the library, there are 73 non-fiction books and 46 fiction books. How many more non-fiction books are there in the library?

Equation: ______________________

Answer: ______________

10) There were 64 bottles of water for sale at a convenience store. After 46 bottles were sold, how many bottles of water are left?

Equation: ______________________

Answer: ______________

65 2 digits − 2 digits ㉕

Date ______

Time spent: min | Score: /

♠ **Subtract.**

(1)
$$\begin{array}{r} 96 \\ -\ 38 \\ \hline \end{array}$$

(2)
$$\begin{array}{r} 83 \\ -\ 47 \\ \hline \end{array}$$

(3)
$$\begin{array}{r} 94 \\ -\ 27 \\ \hline \end{array}$$

(4)
$$\begin{array}{r} 90 \\ -\ 45 \\ \hline \end{array}$$

(5)
$$\begin{array}{r} 81 \\ -\ 19 \\ \hline \end{array}$$

(6)
$$\begin{array}{r} 98 \\ -\ 59 \\ \hline \end{array}$$

(7)
$$\begin{array}{r} 83 \\ -\ 35 \\ \hline \end{array}$$

(8)
$$\begin{array}{r} 92 \\ -\ 76 \\ \hline \end{array}$$

(9) $\begin{array}{r} 80 \\ -\ 26 \\ \hline \end{array}$

(10) $\begin{array}{r} 91 \\ -\ 46 \\ \hline \end{array}$

(11) $\begin{array}{r} 97 \\ -\ 58 \\ \hline \end{array}$

(12) $\begin{array}{r} 91 \\ -\ 63 \\ \hline \end{array}$

(13) $\begin{array}{r} 87 \\ -\ 69 \\ \hline \end{array}$

(14) $\begin{array}{r} 93 \\ -\ 66 \\ \hline \end{array}$

(15) $\begin{array}{r} 81 \\ -\ 48 \\ \hline \end{array}$

(16) $\begin{array}{r} 86 \\ -\ 27 \\ \hline \end{array}$

(17) $\begin{array}{r} 92 \\ -\ 48 \\ \hline \end{array}$

(18) $\begin{array}{r} 95 \\ -\ 29 \\ \hline \end{array}$

66 2 digits − 2 digits ㉖

Date

Time spent: min

Score: /

♠ **Subtract.**

(1) $\begin{array}{r} 95 \\ -\ 16 \\ \hline \end{array}$

(2) $\begin{array}{r} 90 \\ -\ 37 \\ \hline \end{array}$

(3) $\begin{array}{r} 82 \\ -\ 47 \\ \hline \end{array}$

(4) $\begin{array}{r} 98 \\ -\ 29 \\ \hline \end{array}$

(5) $\begin{array}{r} 84 \\ -\ 36 \\ \hline \end{array}$

(6) $\begin{array}{r} 92 \\ -\ 59 \\ \hline \end{array}$

(7) $\begin{array}{r} 97 \\ -\ 39 \\ \hline \end{array}$

(8) $\begin{array}{r} 91 \\ -\ 65 \\ \hline \end{array}$

9) There are 95 books in the kid's section of a library. If 48 books are checked out from the kid's section, how many books are left?

Equation: ______________________________

Answer: __________________

10) I took a math test at school and scored 68 out of 100 last month. I worked very hard on math and scored 87 on the math test this month. How much higher was my math test score this month?

Equation: ______________________________

Answer: __________________

Date

Time spent | Score

min

67 2 digits − 2 digits ㉗

♠ **Subtract.**

(1)
$$\begin{array}{r} 84 \\ -\ 19 \\ \hline \end{array}$$

(2)
$$\begin{array}{r} 80 \\ -\ 55 \\ \hline \end{array}$$

(3)
$$\begin{array}{r} 97 \\ -\ 49 \\ \hline \end{array}$$

(4)
$$\begin{array}{r} 85 \\ -\ 47 \\ \hline \end{array}$$

(5)
$$\begin{array}{r} 94 \\ -\ 38 \\ \hline \end{array}$$

(6)
$$\begin{array}{r} 87 \\ -\ 49 \\ \hline \end{array}$$

(7)
$$\begin{array}{r} 95 \\ -\ 29 \\ \hline \end{array}$$

(8)
$$\begin{array}{r} 82 \\ -\ 37 \\ \hline \end{array}$$

(9)
$$\begin{array}{r} 81 \\ -\ 68 \\ \hline \end{array}$$

(10)
$$\begin{array}{r} 83 \\ -\ 39 \\ \hline \end{array}$$

(11)
$$\begin{array}{r} 82 \\ -\ 49 \\ \hline \end{array}$$

(12)
$$\begin{array}{r} 93 \\ -\ 28 \\ \hline \end{array}$$

(13)
$$\begin{array}{r} 94 \\ -\ 47 \\ \hline \end{array}$$

(14)
$$\begin{array}{r} 91 \\ -\ 56 \\ \hline \end{array}$$

(15)
$$\begin{array}{r} 81 \\ -\ 67 \\ \hline \end{array}$$

(16)
$$\begin{array}{r} 85 \\ -\ 59 \\ \hline \end{array}$$

(17)
$$\begin{array}{r} 93 \\ -\ 38 \\ \hline \end{array}$$

(18)
$$\begin{array}{r} 88 \\ -\ 19 \\ \hline \end{array}$$

68 2 digits − 2 digits ㉘

Date

Time spent | min

Score

♠ **Subtract.**

(1)
$$\begin{array}{r} 96 \\ -\ 28 \\ \hline \end{array}$$

(2)
$$\begin{array}{r} 92 \\ -\ 36 \\ \hline \end{array}$$

(3)
$$\begin{array}{r} 90 \\ -\ 47 \\ \hline \end{array}$$

(4)
$$\begin{array}{r} 84 \\ -\ 39 \\ \hline \end{array}$$

(5)
$$\begin{array}{r} 93 \\ -\ 66 \\ \hline \end{array}$$

(6)
$$\begin{array}{r} 83 \\ -\ 57 \\ \hline \end{array}$$

(7)
$$\begin{array}{r} 84 \\ -\ 68 \\ \hline \end{array}$$

(8)
$$\begin{array}{r} 81 \\ -\ 39 \\ \hline \end{array}$$

9) I read for 58 minutes last week and for 85 minutes this week. How many more minutes did I read this week than last week?

Equation: ______________________________

Answer: __________________

10) In the library, 29 books were checked out on Monday and 54 books on Tuesday. How many more books were checked out on Tuesday?

Equation: ______________________________

Answer: __________________

2 digits − 2 digits ㉙

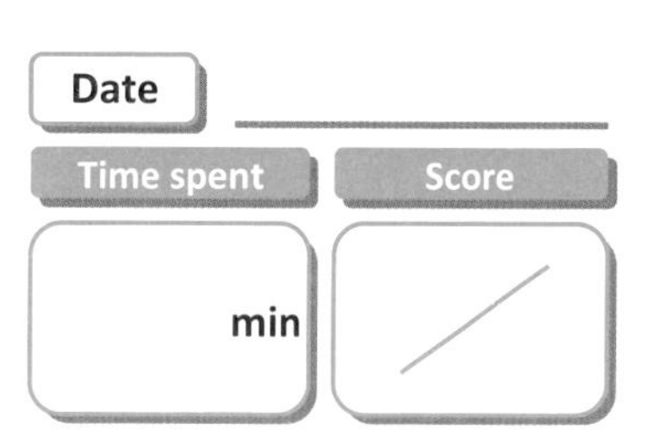

♠ **Subtract.**

(1)
$$\begin{array}{r} 83 \\ -\ 55 \\ \hline \end{array}$$

(2)
$$\begin{array}{r} 91 \\ -\ 65 \\ \hline \end{array}$$

(3)
$$\begin{array}{r} 60 \\ -\ 41 \\ \hline \end{array}$$

(4)
$$\begin{array}{r} 63 \\ -\ 26 \\ \hline \end{array}$$

(5)
$$\begin{array}{r} 82 \\ -\ 28 \\ \hline \end{array}$$

(6)
$$\begin{array}{r} 72 \\ -\ 59 \\ \hline \end{array}$$

(7)
$$\begin{array}{r} 80 \\ -\ 58 \\ \hline \end{array}$$

(8)
$$\begin{array}{r} 65 \\ -\ 37 \\ \hline \end{array}$$

(9) $\begin{array}{r} 80 \\ -\ 36 \\ \hline \end{array}$

(10) $\begin{array}{r} 61 \\ -\ 44 \\ \hline \end{array}$

(11) $\begin{array}{r} 68 \\ -\ 39 \\ \hline \end{array}$

(12) $\begin{array}{r} 82 \\ -\ 27 \\ \hline \end{array}$

(13) $\begin{array}{r} 64 \\ -\ 26 \\ \hline \end{array}$

(14) $\begin{array}{r} 90 \\ -\ 32 \\ \hline \end{array}$

(15) $\begin{array}{r} 85 \\ -\ 56 \\ \hline \end{array}$

(16) $\begin{array}{r} 61 \\ -\ 25 \\ \hline \end{array}$

(17) $\begin{array}{r} 71 \\ -\ 57 \\ \hline \end{array}$

(18) $\begin{array}{r} 94 \\ -\ 46 \\ \hline \end{array}$

70 2 digits − 2 digits ㉚

Date

Time spent: min

Score

♠ **Subtract.**

(1) $\begin{array}{r} 65 \\ -\ 28 \\ \hline \end{array}$

(2) $\begin{array}{r} 64 \\ -\ 36 \\ \hline \end{array}$

(3) $\begin{array}{r} 78 \\ -\ 19 \\ \hline \end{array}$

(4) $\begin{array}{r} 86 \\ -\ 38 \\ \hline \end{array}$

(5) $\begin{array}{r} 73 \\ -\ 47 \\ \hline \end{array}$

(6) $\begin{array}{r} 65 \\ -\ 46 \\ \hline \end{array}$

(7) $\begin{array}{r} 60 \\ -\ 17 \\ \hline \end{array}$

(8) $\begin{array}{r} 61 \\ -\ 28 \\ \hline \end{array}$

9) Last year, Ryan read 90 books. If Logan read 77 books last year, how many more books did Ryan read than Logan last year?

Equation: ______________________________

Answer: __________________

10) There were 85 birds sitting in a tree. After a while, 37 birds flew away. Then how many birds are left in the tree?

Equation: ______________________________

Answer: __________________

Week 4

This week's goal is to practice subtracting 2 digit numbers from 2 digit numbers with or without borrowing.

Tiger Session

Monday	71	72
Tuesday	73	74
Wednesday	75	76
Thursday	77	78
Friday	79	80

71 2 digits − 2 digits ㉛

Date

Time spent min

Score

♠ **Subtract.**

(1) $\begin{array}{r} 71 \\ -\ 65 \\ \hline \end{array}$

(2) $\begin{array}{r} 63 \\ -\ 36 \\ \hline \end{array}$

(3) $\begin{array}{r} 32 \\ -\ 13 \\ \hline \end{array}$

(4) $\begin{array}{r} 27 \\ -\ 12 \\ \hline \end{array}$

(5) $\begin{array}{r} 24 \\ -\ 15 \\ \hline \end{array}$

(6) $\begin{array}{r} 60 \\ -\ 23 \\ \hline \end{array}$

(7) $\begin{array}{r} 86 \\ -\ 37 \\ \hline \end{array}$

(8) $\begin{array}{r} 84 \\ -\ 26 \\ \hline \end{array}$

(9) $\begin{array}{r} 54 \\ -\ 17 \\ \hline \end{array}$

(10) $\begin{array}{r} 46 \\ -\ 18 \\ \hline \end{array}$

(11) $\begin{array}{r} 95 \\ -\ 26 \\ \hline \end{array}$

(12) $\begin{array}{r} 53 \\ -\ 36 \\ \hline \end{array}$

(13) $\begin{array}{r} 49 \\ -\ 29 \\ \hline \end{array}$

(14) $\begin{array}{r} 44 \\ -\ 28 \\ \hline \end{array}$

(15) $\begin{array}{r} 34 \\ -\ 19 \\ \hline \end{array}$

(16) $\begin{array}{r} 67 \\ -\ 28 \\ \hline \end{array}$

(17) $\begin{array}{r} 27 \\ -\ 19 \\ \hline \end{array}$

(18) $\begin{array}{r} 87 \\ -\ 49 \\ \hline \end{array}$

72 2 digits − 2 digits ㉜

Date

Time spent min

Score

♠ **Subtract.**

(1) $\begin{array}{r} 35 \\ -\ 17 \\ \hline \end{array}$

(2) $\begin{array}{r} 82 \\ -\ 45 \\ \hline \end{array}$

(3) $\begin{array}{r} 35 \\ -\ 19 \\ \hline \end{array}$

(4) $\begin{array}{r} 21 \\ -\ 16 \\ \hline \end{array}$

(5) $\begin{array}{r} 71 \\ -\ 38 \\ \hline \end{array}$

(6) $\begin{array}{r} 87 \\ -\ 29 \\ \hline \end{array}$

(7) $\begin{array}{r} 51 \\ -\ 28 \\ \hline \end{array}$

(8) $\begin{array}{r} 53 \\ -\ 27 \\ \hline \end{array}$

9) 41 crocodiles live in a swamp. 33 out of the 41 are adult crocodiles, and the others are baby crocodiles. Then how many baby crocodiles are there in the swamp?

Equation: ______________________

Answer: ______________

10) There were 46 trees in the forest. After some trees were planted, now there are 83 trees in the forest. How many trees were planted?

Equation: ______________________

Answer: ______________

73 2 digits − 2 digits ㉝

Date ______

Time spent | Score
min

♠ **Subtract.**

(1) $\begin{array}{r} 61 \\ -\ 22 \\ \hline \end{array}$

(2) $\begin{array}{r} 30 \\ -\ 12 \\ \hline \end{array}$

(3) $\begin{array}{r} 32 \\ -\ 19 \\ \hline \end{array}$

(4) $\begin{array}{r} 35 \\ -\ 18 \\ \hline \end{array}$

(5) $\begin{array}{r} 45 \\ -\ 17 \\ \hline \end{array}$

(6) $\begin{array}{r} 80 \\ -\ 28 \\ \hline \end{array}$

(7) $\begin{array}{r} 33 \\ -\ 15 \\ \hline \end{array}$

(8) $\begin{array}{r} 72 \\ -\ 27 \\ \hline \end{array}$

(9) $\begin{array}{r} 27 \\ -\ 18 \\ \hline \end{array}$

(10) $\begin{array}{r} 57 \\ -\ 29 \\ \hline \end{array}$

(11) $\begin{array}{r} 77 \\ -\ 39 \\ \hline \end{array}$

(12) $\begin{array}{r} 44 \\ -\ 27 \\ \hline \end{array}$

(13) $\begin{array}{r} 47 \\ -\ 29 \\ \hline \end{array}$

(14) $\begin{array}{r} 34 \\ -\ 19 \\ \hline \end{array}$

(15) $\begin{array}{r} 24 \\ -\ 15 \\ \hline \end{array}$

(16) $\begin{array}{r} 58 \\ -\ 39 \\ \hline \end{array}$

(17) $\begin{array}{r} 54 \\ -\ 25 \\ \hline \end{array}$

(18) $\begin{array}{r} 61 \\ -\ 39 \\ \hline \end{array}$

Date

Time spent | Score

min

74 2 digits − 2 digits ㉞

♠ Subtract.

(1)
$$\begin{array}{r} 45 \\ -\ 28 \\ \hline \end{array}$$

(2)
$$\begin{array}{r} 48 \\ -\ 19 \\ \hline \end{array}$$

(3)
$$\begin{array}{r} 41 \\ -\ 15 \\ \hline \end{array}$$

(4)
$$\begin{array}{r} 21 \\ -\ 16 \\ \hline \end{array}$$

(5)
$$\begin{array}{r} 35 \\ -\ 19 \\ \hline \end{array}$$

(6)
$$\begin{array}{r} 30 \\ -\ 15 \\ \hline \end{array}$$

(7)
$$\begin{array}{r} 61 \\ -\ 29 \\ \hline \end{array}$$

(8)
$$\begin{array}{r} 42 \\ -\ 28 \\ \hline \end{array}$$

9) There are 90 pairs of shoes in a shoe store. If 47 pairs are sold, how many pairs of shoes are left?

Equation: ______________________________

Answer: ____________________

10) There are 49 yellow cups and 67 blue cups on a shelf. How many more blue cups are there?

Equation: ______________________________

Answer: ____________________

Date

Time spent | Score

min

75 2 digits − 2 digits ㉟

♠ Subtract.

(1) $\begin{array}{r} 56 \\ -\ 29 \\ \hline \end{array}$

(2) $\begin{array}{r} 63 \\ -\ 29 \\ \hline \end{array}$

(3) $\begin{array}{r} 71 \\ -\ 12 \\ \hline \end{array}$

(4) $\begin{array}{r} 89 \\ -\ 33 \\ \hline \end{array}$

(5) $\begin{array}{r} 45 \\ -\ 19 \\ \hline \end{array}$

(6) $\begin{array}{r} 32 \\ -\ 16 \\ \hline \end{array}$

(7) $\begin{array}{r} 65 \\ -\ 17 \\ \hline \end{array}$

(8) $\begin{array}{r} 55 \\ -\ 23 \\ \hline \end{array}$

(9) $\begin{array}{r} 33 \\ -\ 28 \\ \hline \end{array}$

(10) $\begin{array}{r} 49 \\ -\ 26 \\ \hline \end{array}$

(11) $\begin{array}{r} 72 \\ -\ 28 \\ \hline \end{array}$

(12) $\begin{array}{r} 79 \\ -\ 37 \\ \hline \end{array}$

(13) $\begin{array}{r} 54 \\ -\ 25 \\ \hline \end{array}$

(14) $\begin{array}{r} 52 \\ -\ 13 \\ \hline \end{array}$

(15) $\begin{array}{r} 54 \\ -\ 27 \\ \hline \end{array}$

(16) $\begin{array}{r} 65 \\ -\ 48 \\ \hline \end{array}$

(17) $\begin{array}{r} 59 \\ -\ 17 \\ \hline \end{array}$

(18) $\begin{array}{r} 40 \\ -\ 24 \\ \hline \end{array}$

Date ______

Time spent: min | Score: /

76 2 digits − 2 digits ㊱

♠ **Subtract.**

(1)
$$\begin{array}{r} 26 \\ -\ 18 \\ \hline \end{array}$$

(2)
$$\begin{array}{r} 47 \\ -\ 29 \\ \hline \end{array}$$

(3)
$$\begin{array}{r} 28 \\ -\ 13 \\ \hline \end{array}$$

(4)
$$\begin{array}{r} 54 \\ -\ 17 \\ \hline \end{array}$$

(5)
$$\begin{array}{r} 42 \\ -\ 24 \\ \hline \end{array}$$

(6)
$$\begin{array}{r} 80 \\ -\ 71 \\ \hline \end{array}$$

(7)
$$\begin{array}{r} 69 \\ -\ 29 \\ \hline \end{array}$$

(8)
$$\begin{array}{r} 41 \\ -\ 27 \\ \hline \end{array}$$

9) There are 65 pants and 35 t-shirts in a store. How many more pants are in the store than the t-shirts?

Equation: ____________________

Answer: ____________

10) In a town, there were 76 houses last year. This year, there are 92 houses. How many more houses have been built in the town between last year and this year?

Equation: ____________________

Answer: ____________

Date

Time spent | Score

min

77 2 digits − 2 digits ㊲

♠ **Subtract.**

(1) $\begin{array}{r} 81 \\ -\ 16 \\ \hline \end{array}$

(2) $\begin{array}{r} 32 \\ -\ 17 \\ \hline \end{array}$

(3) $\begin{array}{r} 24 \\ -\ 18 \\ \hline \end{array}$

(4) $\begin{array}{r} 27 \\ -\ 19 \\ \hline \end{array}$

(5) $\begin{array}{r} 78 \\ -\ 30 \\ \hline \end{array}$

(6) $\begin{array}{r} 50 \\ -\ 25 \\ \hline \end{array}$

(7) $\begin{array}{r} 43 \\ -\ 27 \\ \hline \end{array}$

(8) $\begin{array}{r} 90 \\ -\ 39 \\ \hline \end{array}$

(9) $\begin{array}{r} 46 \\ -\ 28 \\ \hline \end{array}$

(10) $\begin{array}{r} 56 \\ -\ 29 \\ \hline \end{array}$

(11) $\begin{array}{r} 43 \\ -\ 19 \\ \hline \end{array}$

(12) $\begin{array}{r} 25 \\ -\ 17 \\ \hline \end{array}$

(13) $\begin{array}{r} 78 \\ -\ 59 \\ \hline \end{array}$

(14) $\begin{array}{r} 24 \\ -\ 16 \\ \hline \end{array}$

(15) $\begin{array}{r} 71 \\ -\ 48 \\ \hline \end{array}$

(16) $\begin{array}{r} 67 \\ -\ 38 \\ \hline \end{array}$

(17) $\begin{array}{r} 32 \\ -\ 19 \\ \hline \end{array}$

(18) $\begin{array}{r} 85 \\ -\ 27 \\ \hline \end{array}$

Date ______

Time spent | Score

min

78 2 digits − 2 digits ㊳

♠ **Subtract.**

(1)
$$\begin{array}{r} 54 \\ -\ 16 \\ \hline \end{array}$$

(2)
$$\begin{array}{r} 81 \\ -\ 24 \\ \hline \end{array}$$

(3)
$$\begin{array}{r} 28 \\ -\ 13 \\ \hline \end{array}$$

(4)
$$\begin{array}{r} 34 \\ -\ 19 \\ \hline \end{array}$$

(5)
$$\begin{array}{r} 44 \\ -\ 17 \\ \hline \end{array}$$

(6)
$$\begin{array}{r} 27 \\ -\ 18 \\ \hline \end{array}$$

(7)
$$\begin{array}{r} 64 \\ -\ 27 \\ \hline \end{array}$$

(8)
$$\begin{array}{r} 84 \\ -\ 29 \\ \hline \end{array}$$

9) Cole's family put 43 items out for sale in a yard sale. If 37 items were sold, how many items are left?

Equation: ______________________

Answer: ______________

10) Michelle's family did a garage sale for two days in a row. They earned 56 dollars for the first day and 93 dollars for the second day. How many more dollars did they make in the second day?

Equation: ______________________

Answer: ______________

79 2 digits − 2 digits ㊴

Date

Time spent min

Score

♠ **Subtract.**

(1) $\begin{array}{r} 49 \\ -\ 23 \\ \hline \end{array}$

(2) $\begin{array}{r} 80 \\ -\ 24 \\ \hline \end{array}$

(3) $\begin{array}{r} 33 \\ -\ 25 \\ \hline \end{array}$

(4) $\begin{array}{r} 61 \\ -\ 33 \\ \hline \end{array}$

(5) $\begin{array}{r} 89 \\ -\ 42 \\ \hline \end{array}$

(6) $\begin{array}{r} 70 \\ -\ 61 \\ \hline \end{array}$

(7) $\begin{array}{r} 59 \\ -\ 21 \\ \hline \end{array}$

(8) $\begin{array}{r} 48 \\ -\ 14 \\ \hline \end{array}$

(9) $\begin{array}{r} 45 \\ -\ 27 \\ \hline \end{array}$

(10) $\begin{array}{r} 33 \\ -\ 15 \\ \hline \end{array}$

(11) $\begin{array}{r} 72 \\ -\ 26 \\ \hline \end{array}$

(12) $\begin{array}{r} 82 \\ -\ 35 \\ \hline \end{array}$

(13) $\begin{array}{r} 75 \\ -\ 37 \\ \hline \end{array}$

(14) $\begin{array}{r} 52 \\ -\ 18 \\ \hline \end{array}$

(15) $\begin{array}{r} 69 \\ -\ 25 \\ \hline \end{array}$

(16) $\begin{array}{r} 42 \\ -\ 15 \\ \hline \end{array}$

(17) $\begin{array}{r} 45 \\ -\ 26 \\ \hline \end{array}$

(18) $\begin{array}{r} 31 \\ -\ 18 \\ \hline \end{array}$

80 2 digits − 2 digits ㊵

Date

Time spent	Score
min	/

♠ **Subtract.**

(1) $\begin{array}{r} 59 \\ -\ 22 \\ \hline \end{array}$

(2) $\begin{array}{r} 48 \\ -\ 16 \\ \hline \end{array}$

(3) $\begin{array}{r} 74 \\ -\ 49 \\ \hline \end{array}$

(4) $\begin{array}{r} 34 \\ -\ 16 \\ \hline \end{array}$

(5) $\begin{array}{r} 22 \\ -\ 11 \\ \hline \end{array}$

(6) $\begin{array}{r} 52 \\ -\ 29 \\ \hline \end{array}$

(7) $\begin{array}{r} 91 \\ -\ 36 \\ \hline \end{array}$

(8) $\begin{array}{r} 37 \\ -\ 19 \\ \hline \end{array}$

9) My family decided to donate gift boxes to 50 children. If we made 35 boxes yesterday, how many more boxes do we need to make?

Equation: ______________________________

Answer: __________________

10) My family donated 57 books to the library last year and 81 books this year. How many more books did my family donate this year than last year?

Equation: ______________________________

Answer: __________________

D – 2: Answers

Week 1

41	(p. 5 ~ 6)			
① 24	② 17	③ 11	④ 23	⑤ 14
⑥ 12	⑦ 11	⑧ 16	⑨ 22	⑩ 20
⑪ 16	⑫ 15	⑬ 12	⑭ 17	⑮ 13
⑯ 14				

42	(p. 7 ~ 8)			
① 25	② 14	③ 21	④ 20	⑤ 16
⑥ 23	⑦ 13	⑧ 12		
⑨ 23 – 12 = 11, 11 more notebooks				
⑩ 36 – 13 = 23, 23 more apples				

43	(p. 9 ~ 10)			
① 32	② 21	③ 3	④ 44	⑤ 14
⑥ 21	⑦ 30	⑧ 41	⑨ 40	⑩ 24
⑪ 10	⑫ 50	⑬ 40	⑭ 13	⑮ 14
⑯ 31	⑰ 32	⑱ 12		

44	(p. 11 ~ 12)			
① 41	② 21	③ 20	④ 30	⑤ 34
⑥ 15	⑦ 14	⑧ 36		
⑨ 56 – 32 = 24, 24 dolls				
⑩ 48 – 12 = 36, 36 robots				

45	(p. 13 ~ 14)			
① 53	② 47	③ 25	④ 44	⑤ 14
⑥ 32	⑦ 22	⑧ 43	⑨ 20	⑩ 22
⑪ 23	⑫ 61	⑬ 48	⑭ 34	⑮ 13
⑯ 50	⑰ 40	⑱ 14		

46	(p. 15 ~ 16)			
① 11	② 40	③ 51	④ 16	⑤ 23
⑥ 60	⑦ 31	⑧ 14		
⑨ 77 – 56 = 21, 21 more times				
⑩ 65 – 32 = 33, 33 pencils				

47	(p. 17 ~ 18)			
① 13	② 31	③ 65	④ 32	⑤ 52
⑥ 46	⑦ 44	⑧ 37	⑨ 11	⑩ 47
⑪ 40	⑫ 22	⑬ 23	⑭ 35	⑮ 41
⑯ 60	⑰ 20	⑱ 26		

48	(p. 19 ~ 20)			
① 33	② 25	③ 28	④ 30	⑤ 31
⑥ 23	⑦ 24	⑧ 11		
⑨ 98 – 56 = 42, 42 pizzas				
⑩ 57 – 45 = 12, 12 more laps				

49	(p. 21 ~ 22)			
① 21	② 44	③ 34	④ 23	⑤ 22
⑥ 25	⑦ 36	⑧ 14	⑨ 46	⑩ 13
⑪ 31	⑫ 45	⑬ 50	⑭ 22	⑮ 33
⑯ 15	⑰ 31	⑱ 43		

50	(p. 23 ~ 24)			
① 48	② 53	③ 13	④ 11	⑤ 15
⑥ 22	⑦ 32	⑧ 33		
⑨ 65 – 22 = 43, 43 more vans				
⑩ 77 – 37 = 40, 40 years older				

Week 2

51	(p. 27 ~ 28)	
① 2, 17, 19	② 1, 12, 3	③ 2, 11, 15
④ 2, 12, 14	⑤ 2, 15, 16	⑥ 1, 14, 8
⑦ 1, 13, 9	⑧ 1, 17, 18	⑨ 2, 10, 15
⑩ 1, 15, 7		

52	(p. 29 ~ 30)	
① 1, 12, 6	② 1, 17, 9	③ 2, 14, 17
④ 2, 13, 15	⑤ 2, 11, 14	⑥ 1, 11, 6
⑦ 2, 17, 18	⑧ 2, 12, 4	
⑨ 35 – 29 = 6, 6 years older		
⑩ 33 – 18 = 15, 15 pebbles		

53	(p. 31 ~ 32)	
① 2, 13, 16	② 2, 14, 16	③ 2, 13, 18
④ 2, 15, 19	⑤ 1, 13, 9	⑥ 1, 12, 4
⑦ 1, 16, 7	⑧ 2, 11, 16	⑨ 1, 11, 3
⑩ 2, 15, 8	⑪ 1, 14, 5	⑫ 1, 18, 9
⑬ 1, 13, 9	⑭ 2, 16, 17	⑮ 2, 11, 14
⑯ 1, 15, 7	⑰ 2, 12, 16	⑱ 2, 17, 18

54	(p. 33 ~ 34)	
① 1, 15, 8	② 2, 16, 7	③ 2, 18, 19
④ 2, 10, 15	⑤ 1, 11, 6	⑥ 2, 18, 19

⑦ 2, 17, 18	⑧ 1, 12, 4	
⑨ 36 – 19 = 17, 17 more laps		
⑩ 24 – 15 = 9, 9 bottles		

55	(p. 35 ~ 36)	
① 3, 14, 29	② 4, 10, 33	③ 3, 17, 19
④ 4, 17, 28	⑤ 4, 11, 18	⑥ 3, 10, 16
⑦ 3, 16, 28	⑧ 3, 18, 19	⑨ 3, 10, 15
⑩ 4, 18, 29	⑪ 4, 14, 36	⑫ 4, 10, 33
⑬ 4, 11, 24	⑭ 3, 11, 15	⑮ 4, 14, 36
⑯ 4, 17, 28	⑰ 3, 10, 23	⑱ 4, 15, 29

56	(p. 37 ~ 38)	
① 4, 15, 18	② 3, 11, 25	③ 3, 11, 29
④ 3, 16, 17	⑤ 3, 17, 9	⑥ 3, 10, 21
⑦ 3, 12, 15	⑧ 4, 18, 39	
⑨ 52 – 29 = 23, 23 more pants		
⑩ 45 – 26 = 19, 19 shells		

57	(p. 39 ~ 40)	
① 3, 16, 28	② 3, 10, 17	③ 3, 17, 18
④ 3, 11, 23	⑤ 3, 13, 26	⑥ 3, 17, 28
⑦ 4, 17, 39	⑧ 3, 14, 15	⑨ 3, 12, 27
⑩ 3, 13, 27	⑪ 3, 11, 24	⑫ 3, 16, 18
⑬ 3, 13, 14	⑭ 4, 18, 19	⑮ 3, 15, 7
⑯ 4, 10, 19	⑰ 4, 13, 17	⑱ 3, 17, 18

58	(p. 41 ~ 42)	
① 4, 13, 14	② 3, 17, 19	③ 4, 15, 26
④ 4, 10, 13	⑤ 3, 16, 28	⑥ 4, 14, 27
⑦ 3, 15, 16	⑧ 3, 14, 25	
⑨ 32 – 26 = 6, 6 more kittens		
⑩ 52 – 34 = 18, 18 more kids		

59	(p. 43 ~ 44)	
① 1, 15, 9	② 1, 11, 5	③ 3, 17, 29
④ 1, 15, 6	⑤ 3, 14, 29	⑥ 2, 13, 17
⑦ 3, 10, 14	⑧ 1, 14, 7	⑨ 4, 18, 39
⑩ 1, 10, 8	⑪ 1, 14, 9	⑫ 2, 16, 17
⑬ 4, 10, 39	⑭ 1, 13, 6	⑮ 3, 15, 26
⑯ 2, 14, 19	⑰ 2, 16, 17	⑱ 1, 17, 9

60	(p. 45 ~ 46)	
① 2, 11, 15	② 4, 14, 27	③ 2, 15, 7
④ 2, 16, 19	⑤ 4, 12, 13	⑥ 3, 10, 12
⑦ 2, 16, 18	⑧ 3, 16, 27	
⑨ 42 – 35 = 7, 7 pieces		
⑩ 30 – 15 = 15, 15 more stickers		

Week 3

61	(p. 49 ~ 50)			
① 5, 14, 38		② 6, 17, 49		③ 26
④ 16	⑤ 17	⑥ 24	⑦ 45	⑧ 53
⑨ 36	⑩ 26	⑪ 42	⑫ 38	⑬ 15
⑭ 29	⑮ 46	⑯ 55	⑰ 57	⑱ 48

62	(p. 51 ~ 52)			
① 5, 15, 37		② 6, 14, 57		③ 8
④ 18	⑤ 35	⑥ 26	⑦ 49	⑧ 45
⑨ 73 – 68 = 5, 5 more boys				
⑩ 61 – 45 = 16, 16 students				

63	(p. 53 ~ 54)			
① 5, 10, 25		② 5, 11, 42		③ 39
④ 37	⑤ 29	⑥ 36	⑦ 19	⑧ 42
⑨ 44	⑩ 37	⑪ 36	⑫ 35	⑬ 22
⑭ 11	⑮ 23	⑯ 39	⑰ 25	⑱ 18

64	(p. 55 ~ 56)			
① 6, 16, 48		② 5, 10, 24		③ 39
④ 36	⑤ 33	⑥ 14	⑦ 18	⑧ 46
⑨ 73 – 46 = 27, 27 more non-fiction books				
⑩ 64 – 46 = 18, 18 bottles				

65	(p. 57 ~ 58)			
① 8, 16, 58		② 7, 13, 36		③ 67
④ 45	⑤ 62	⑥ 39	⑦ 48	⑧ 16
⑨ 54	⑩ 45	⑪ 39	⑫ 28	⑬ 18
⑭ 27	⑮ 33	⑯ 59	⑰ 44	⑱ 66

66	(p. 59 ~ 60)			
① 8, 15, 79		② 8, 10, 53		③ 35
④ 69	⑤ 48	⑥ 33	⑦ 58	⑧ 26
⑨ 95 – 48 = 47, 47 books				
⑩ 87 – 68 = 19, 19 points higher				

67	(p. 61 ~ 62)			
① 7, 14, 65		② 7, 10, 25		③ 48
④ 38	⑤ 56	⑥ 38	⑦ 66	⑧ 45
⑨ 13	⑩ 44	⑪ 33	⑫ 65	⑬ 47
⑭ 35	⑮ 14	⑯ 26	⑰ 55	⑱ 69

68	(p. 63 ~ 64)			
① 8, 16, 68		② 8, 12, 56		③ 43
④ 45	⑤ 27	⑥ 26	⑦ 16	⑧ 42
⑨ 85 – 58 = 27, 27 more minutes				
⑩ 54 – 29 = 25, 25 more books				

69	(p. 65 ~ 66)			
① 7, 13, 28		② 8, 11, 26		③ 19
④ 37	⑤ 54	⑥ 13	⑦ 22	⑧ 28
⑨ 44	⑩ 17	⑪ 29	⑫ 55	⑬ 38
⑭ 58	⑮ 29	⑯ 36	⑰ 14	⑱ 48

70	(p. 67 ~ 68)			
① 5, 15, 37		② 5, 14, 28		③ 59
④ 48	⑤ 26	⑥ 19	⑦ 43	⑧ 33
⑨ 90 – 77 = 13, 13 more books				
⑩ 85 – 37 = 48, 48 birds				

Week 4

71	(p. 71 ~ 72)			
① 6	② 27	③ 19	④ 15	⑤ 9
⑥ 37	⑦ 49	⑧ 58	⑨ 37	⑩ 28
⑪ 69	⑫ 17	⑬ 20	⑭ 16	⑮ 15
⑯ 39	⑰ 8	⑱ 38		

72	(p. 73 ~ 74)			
① 18	② 37	③ 16	④ 5	⑤ 33
⑥ 58	⑦ 23	⑧ 26		
⑨ 41 – 33 = 8, 8 baby crocodiles				
⑩ 83 – 46 = 37, 37 trees				

73	(p. 75 ~ 76)			
① 39	② 18	③ 13	④ 17	⑤ 28
⑥ 52	⑦ 18	⑧ 45	⑨ 9	⑩ 28
⑪ 38	⑫ 17	⑬ 18	⑭ 15	⑮ 9
⑯ 19	⑰ 29	⑱ 22		

74	(p. 77 ~ 78)			
① 17	② 29	③ 26	④ 5	⑤ 16
⑥ 15	⑦ 32	⑧ 14		
⑨ 90 – 47 = 43, 43 pairs				
⑩ 67 – 49 = 18, 18 more blue cups				

75	(p. 79 ~ 80)			
① 27	② 34	③ 59	④ 56	⑤ 26
⑥ 16	⑦ 48	⑧ 32	⑨ 5	⑩ 23
⑪ 44	⑫ 42	⑬ 29	⑭ 39	⑮ 27
⑯ 17	⑰ 42	⑱ 16		

76	(p. 81 ~ 82)			
① 8	② 18	③ 15	④ 37	⑤ 18
⑥ 9	⑦ 40	⑧ 14		
⑨ 65 – 35 = 30, 30 more pants				
⑩ 92 – 76 = 16, 16 more houses				

77	(p. 83 ~ 84)			
① 65	② 15	③ 6	④ 8	⑤ 48
⑥ 25	⑦ 16	⑧ 51	⑨ 18	⑩ 27
⑪ 24	⑫ 8	⑬ 19	⑭ 8	⑮ 23
⑯ 29	⑰ 13	⑱ 58		

78	(p. 85 ~ 86)			
① 38	② 57	③ 15	④ 15	⑤ 27
⑥ 9	⑦ 37	⑧ 55		
⑨ 43 – 37 = 6, 6 items				
⑩ 93 – 56 = 37, 37 more dollars				

79	(p. 87 ~ 88)			
① 26	② 56	③ 8	④ 28	⑤ 47
⑥ 9	⑦ 38	⑧ 34	⑨ 18	⑩ 18
⑪ 46	⑫ 47	⑬ 38	⑭ 34	⑮ 44
⑯ 27	⑰ 19	⑱ 13		

80	(p. 89 ~ 90)			
① 37	② 32	③ 25	④ 18	⑤ 11
⑥ 23	⑦ 55	⑧ 18		
⑨ 50 – 35 = 15, 15 more boxes				
⑩ 81 – 57 = 24, 24 more books				

Tiger Math

ACHIEVEMENT AWARD

THIS AWARD IS PRESENTED TO

____

(student name)

FOR SUCESSFULLY COMPLETING

TIGER MATH LEVEL D – 2.

Dr. Tiger

____

Dr.Tiger